ON THE ASTROLABE

JOHN PHILOPONUS OF ALEXANDRIA

TREATISE CONCERNING THE USING AND ARRANGEMENT OF THE ASTROLABE AND THE THINGS ENGRAVED UPON IT: THAT IS TO SAY, WHAT EACH SIGNIFIES.

JOHN PHILOPONUS OF ALEXANDRIA

[Translated by Herbert Wilson Greene, d. 1933]

THE explanation of the surface of the sphere, and the causes of the things engraved upon it, its use, and the number and nature of the purposes for which it is serviceable, I will try to set out as clearly as I can. The subject has already been treated sufficiently by my teacher the philosopher Ammonius, but still requires to be further elucidated so that it may be easily apprehended by those also who are not instructed in such matters. This some of my friends have urged me to do. First, then, I will state what each of the things engraved upon it is.

1. *Concerning the Engraving on the Plane Surface, on which the Rule lies, and what is signified by each of the things engraved upon it.*

The two straight lines on the plane surface, on which the rule lies, cutting one another in the middle, correspond to the meridian line and the horizon: of these, the one descending from above from the ring whereby we suspend the instrument corresponds to the meridian line in each zone, the other cutting it in two at right angles corresponds to the horizon. On this latter, the one corresponding to the horizon, stands a semi-circle having this line itself as its diameter. This semi-circle corresponds to the hemisphere of the heaven above the earth. This semi-circle is cut in two by the other line, that descending from the ring, which we said corresponded to the meridian line, the cutting taking place by the upper end of the line near the ring. Each of the quadrants on either side has been divided into 90 degrees, on which the index of the rule falls, and thereby we determine the meridian height above the horizon of the sun, or any other star, how many degrees each hour it has been raised above the rising or the setting horizon. The ninetieth degree indicates the Sign at the summit in the case of each House, and the first [degree] what is next to the horizon itself, whether the rising or the setting one, as the using of the instrument will teach us as we proceed. It is not the case, however, that in all astrolabes both quadrants have been divided into the ninety degrees, but only one of them, for one, whichever it happens to be, is enough for our observing. For it is possible to know by either [quadrant] how far the sun, or any other star, has been raised from the setting or from the rising horizon. But that we may find it easy to observe when the instrument is suspended with either hand, in some cases both the quadrants have been engraved.

2. *Concerning the Engraving on the Tablets, on which the Zones have been engraved: to what each of the things engraved corresponds, and the Number of Degrees of the Obliquity of the Zodiac.*

The engraving then of the plane surface, on which the rule lies, has been ordered in this fashion: that of the tablets, on which the zones have been engraved, is as follows. On each plane surface of a tablet there are again two straight lines similarly cutting one another, one of which, coming through as from the ring downward, corresponds again to the meridian line, and the other to the horizon. For they are the same as those on the opposite side, on which the rule lies. It will therefore be necessary to fit the former on to the latter in the same line. There are also circles engraved on the upper part of the tablet, that next to the suspending ring: in the complete astrolabes ninety, in the bipartite forty-five, in the tripartite thirty, or as the engravers please: of these, the outer and larger corresponds to the horizon, and, if it were possible to extend the circle, it would fit on to the straight line cutting the meridian line. Since, however, this is impossible, it naturally results that, being curved, as great a distance of it as falls below at the middle of the straight line, so much on either side it has been raised above its ends. But the straight line as on a plane surface separates the hemisphere above the earth from that beneath it, but the circle does so as on a sphere. The inner and contained circles are parallel to the horizon, standing apart from one another from the horizon itself to the upper one which corresponds to that above the earth, in complete astrolabes by one degree, in the bipartite and tripartite ones by two or three [degrees], so that the hemisphere above the earth is cut by them after the fashion of a crown, in a placing similar to that occupied by the parallel circles in the millstone-like placing of the whole. Hence the inner ones, higher than those near the horizon, are always necessarily smaller, inasmuch as they cut off a lesser rim of the hemisphere above the earth. On the sphere are drawn the said circles, to which those on the instrument correspond, as regards the central point, to the Sign at the summit of each House, as regards distance, the horizon to that coming through from that at the summit to that about the diameter of the whole, the others ever in succession to the distance of this, we further removing either one degree, as in the case of the complete astrolabes, or two or three, as in those of the bipartite and tripartite ones. It is plain that as this distance is of ninety degrees, for it has only the rim of a quadrant, the removal only takes place up to the point where the distance from that at the summit amounts to one degree, as in the case of the complete astrolabes, or to two or more, as in the other cases. The intermediate sign then of the circles, on which has been inscribed the ninetieth degree, corresponds to that at the summit of each house, so that this sign is equivalent to that at the upper end of the line in the other tablet, on which the rule lies, that by the suspending ring itself. For on each the same number that of 90 degrees lies. The meridian line, to which we said the line from the ring corresponded, coming down through the circles themselves, cuts these circles in two, so that the left-hand semi-circles, when the instrument lies facing us, are the rising ones, on which Rising has been inscribed, which the sun or each of the fixed stars moving from the Rising to the meridian, now one now another, touches. Those on the right are the setting ones, on which again Setting has been inscribed, which it [i.e. the sun] touches being borne from the meridian to the Setting. It is plain that owing to the shortness of the instrument not all the circles are complete, but the outer and larger ones falling over outside the circumference of the tablet are only half completed. There lies on the circles their number, from the first to the ninetieth, for the distance from the horizon to that at the summit is, as I said, of so many degrees. And further, let this be clear, that the beginning of the counting is from the horizon, the same numbers having been written upon each of their semicircles, the rising and the setting, as in the case of the outer and half-completed ones. For in the case of the inner and completed ones the number of the circles has been ranged along the line of the meridian line. It is clear, I think, that in the case of bipartite and tripartite astrolabes the distance between the circles is cut into those that have been left on one side. These circles are equivalent to the degrees written on the quadrant of the plane surface on which the rule lies, which we discussed at the beginning.

The semi-circle then of the tablet on which the aforesaid circles were engraved corresponds to the hemisphere above the earth, as the remainder does to that beneath it, which has been divided into twelve sections according to the number of the twelve hours, which the sun makes being in either hemisphere, in that above or in that below the earth. On these lines is laid [inscribed] the number of the hours, the first hour beginning from the setting part for a reason we will state further on.

There are yet three other circles engraved on the aforesaid parallel circles, cutting them and containing one another. Of these the inner one corresponds to the circle of Cancer. When then the Arachne is brought round, you will see the first degree of Cancer, in which the sun makes its summer tropic drawing this circle. Whence the part of it above the earth is greater, that is the part of it borne through the parallel circles: that beneath the earth is smaller, that is what (is borne) through the remaining part of the tablet, on which the lines marking the hours have been impressed, and which, as we have said, corresponds to the hemisphere beneath the earth. The circle second to this, which immediately contains it, corresponds to the Equinoctial, whence the two equinoctial signs, the beginning of Aries and that of Libra, pass through it, and both its semi-circles are equal, that which is drawn through the parallels which is the one above the earth, and that through the lines marking the hours, which indicates what is beneath it. Of these two circles only the semi-circles beneath the earth have been impressed on some instruments, the rest which ought to be borne through the parallels are left to the imagination because the engravings of the parallels are not cut by them. The third which contains them both is in a line with the circle of Capricorn, whence the beginning of Capricorn, in which the winter tropic takes place, passes through this. Consequently the portion of this circle above the earth, i.e. that drawn through the parallels, is smaller, and that beneath the earth, i.e. that drawn through the lines indicating the hours, is greater. Of these three circles, I mean, those of Cancer, of the Equinox, and of Capricorn, the first of the parallels separates in each case the section above the earth from that beneath it, since we said that it corresponds to the horizon. From the circle of Capricorn to that of Cancer the width of the zodiac is forty-seven degrees, narrow first forty-eight and second forty. The distance from the Circle of Capricorn to that of Cancer is forty-eight degrees, as we can know from the inscription on the parallels. For that of Cancer is distant towards the north from the equinoctial twenty-four degrees, as that of Capricorn is towards the south another twenty-four degrees. For having adjusted the Arachne to any one of the zones, and marked the parallel where the beginning of Capricorn touches it at the meridian line, and secondly again where those of Aries and Libra do so, and thirdly where that of Cancer does so, and counted up the parallels in between, you will find from Capricorn to Aries and Libra twenty-four parallels, and from them to Cancer twenty-four more, so that the distance from Capricorn to Cancer is forty-eight degrees, over which distance the obliquity of the zodiac extends. The zone has also been inscribed for which the engraving has been made on each plane surface, as has also the number of equal hours of which the longest day consists in that zone, and of the degrees which the said zone is distant from the equinoctial. It is obvious that the north pole has been raised above the horizon by the same number of degrees as the south pole is distant from it beneath the earth. For it is plain that by as much as each house is distant from the equinoctial, by so much is the north pole raised above the horizon, and the south pole removed from it beneath the earth. In some astrolabes, and especially in the complete ones, the plane itself too, on which the rule lies, has been engraved for each one of the zones. In some of them the outer edge has been divided into three hundred and sixty degrees.

3. *Concerning the things engraved on the Arachne.*

So much then for the tablets and the meaning (or, object) of each of the things engraved upon them. The Arachne which is laid upon them has the zodiac and some of the brighter of the fixed stars. The complete circle on it beginning from outside is the zodiac, the others, which are half completed, contain certain of the fixed stars, of which we will speak in due season. On the zodiac have been engraved its twelve signs from Aries to Pisces. In the complete instruments each sign of the zodiac has been divided into thirty degrees, in the bipartite into fifteen and in the tripartite, as is obvious, into ten, as was the case in the drawing of the parallels. The beginning of the degrees of each sign of the zodiac is at the part where its first line has been drawn, or, to put it otherwise, at the part where is the sign of the zodiac preceding it. For example, the sign of the zodiac preceding Aries is Pisces: the beginning then of Aries is from the part next Pisces, and so in every case. Of the lines that indicate the degrees: some come right through the width of the zodiac, others only half-way: the beginning of each sign of the Zodiac is from the line which comes right through, for that is the end of the preceding sign of the zodiac, and the beginning of the following one.

This then is the arrangement of the whole instrument, and now is the proper time to go through in detail what concerns the using of it.

4. *Concerning the Observation of the Sun by Day, and how we may set about it according to Rule.*

If then we wish to take the hour of the sun by the instrument in the daytime, we suspend the instrument from the ring in such a way that its quadrant, the one cut up into the 90 degrees, inclines towards the sun, and afterwards bring round the rule little by little above and below after the said one and the same quadrant of the central point, until the ray entering through the hole of the rule facing the sun falls upon the other facing us. In order that we may not, by handling the instrument without rational method, find ourselves in difficulty with the rule, it is necessary to know that the instrument must be so placed that its outer edge, I mean the circumference, is shone upon by the sun, and each of the plane surfaces, as far as possible, is in the shade. The reason is this, that the sign on the suspending ring corresponds to the pole of the horizon, that is to say, the sign at the summit, while the circumference of the instrument corresponds to the parallel which the sun draws then being observed. It is necessary then that the circumference should lie so as to be on the same plane surface as the parallel which the sun then draws. And this will be the case if the rays of the sun themselves strike exactly on the edge of the instrument, as if the star (i.e. the sun) were lying upon it. The instrument then having been placed in this position, it is necessary, as I said, to bring the rule round gently above and below to one and the same quadrant of the engraved semi-circle, that inclining towards the sun, until the rule having become in a straight line with the sun the ray of the sun having come through the hole of the rule so ordered as to face it passes through to the other [hole] ordered to face us. While the rule is being brought round, you will see a light of equal size and similar shape to the hole wandering about, and going now in this direction, now in that, with the movement of the rule. It is necessary then to bring round the rule gently, in this direction and in that, until we see the light introduced on to the inner plane surface of the sight-vane on our side, and fitting into its hole, when finally it happens that it becomes invisible as passing through vacuity. If at anyrate you bring your hand near the hole on our side you will see the light falling upon it. It happens that the light becomes altogether invisible if the hole by which it first enters is less than the other or exactly equal. For if it were found to be greater, it happens that the light falls outside the other upon the inner plane surface of the sight-vane facing us. This done, it is necessary to mark with ink or something of the sort the line on which the index of the rule fell (i.e. the end of the little rod terminating in a sharp point) and to measure how great it is, beginning from below from the horizon, whether the observation is taken before the meridian or after it. For as many as are the degrees from the horizon, so great is the exaltation of the sun from the Rising or the Setting. Having marked then the degree, on which the sun is being observed, as, should it so happen, the thirtieth, it is necessary to take from the Calendar the sign of the zodiac and its degree, in which the sun is on that day of which we wish to find the hour, or by means of the method we are going to mention next.

Example.

Let it happen to be in Aries in the twentieth degree. It is necessary then to mark with ink or wax, or something of the sort, the twentieth degree of Aries upon the zodiac on the Arachne, then next to look in what zone we happen to be while observing, and to take the tablet, on which the said zone has been engraved, and so adjust it to the present instrument that the zone we are seeking is outside of all.[5] Then to place the Arachne upon the instrument, and if the observation is taken before midday we must take the parallel circle on the tablet of the said zone equal in number to the observed degree, as now, in supposition, the thirtieth, making the beginning of the counting from the part on which Rising has been inscribed, but if we make the beginning after midday, from the opposite part on which has been inscribed Setting. Then it is necessary to mark this circle with ink with numerous dots along nearly all the line. If, however, the astrolabe is not complete, but bipartite or tripartite, and the observed number of the degrees falls in the space between the circles, it is obviously necessary to cut the intervening distance correspondently, and to mark similarly with numerous dots from above to below the place where falls the number we are seeking. This done we must bring round the Arachne until the sign of the zodiac and the degree in it in which the sun is touch the parallel circle on which the sun is observed to be, and which we directed to be marked by numerous dots because it was uncertain which of them the degree of the sun will touch when the Arachne is brought round. This done, it is necessary to know that the instrument has been adjusted in the same position as that of the whole (? the universe) at that hour and in the same line with the whole. After this it is necessary to take the degree of the sun according to diameter, as, in the present case, the 20th of Libra, and mark with ink on which sign of the tablet it has fallen. It falls altogether in that corresponding to the part of it beneath the earth; then counting thus the lines indicating the hours from the first of the setting, which makes the beginning from the setting part, declare the hours of the sun that have been completed, or the portion, if the degree of the sun according to diameter does not fall upon one of the lines marking the hours but in the intervening distance. And so in the afternoon observation. The only difference is this, that in the taking of the parallel circles in the observations before midday we begin the counting from the rising, and in those after it from the setting. In the taking of the hours, however, we always begin from the setting part, whether the observation is by day or by night, for the reason that we will now state.

5. *Why the Lines marking the Hours are engraved on the section corresponding to that beneath the Earth, and why we begin counting them, and how the Part of an Hour can be ascertained.* Since Ptolemaeus always took the utmost pains to secure clearness and facility, and knew that if he made his engraving in that [part] corresponding to the hemisphere above the earth (in which he made the engraving of the parallel circles) he would have made confusion in the instrument, and made it difficult for users to distinguish which lines indicated the hours, and which the parallels, for this reason he has engraved the hours on the other semicircle. Then it being plain that as great as is the part of a circle above the earth, which the sun passes through in each degree, so great is that beneath the earth which the degree of the sun according to diameter draws (as, for example, as great a part of a circle as the twentieth degree of Aries draws above the earth, so great beneath the earth does the twentieth of Libra, and so in all those according to diameter), and that as far as the sun when above the earth is distant from the rising horizon, so much is its degree according to diameter beneath the earth from the setting horizon, it therefore makes no difference for the knowing of the amount of the distance of the sun from the Rising whether one divides this itself, or the degree according to diameter beneath the earth from the setting horizon. For, as we said, it has been shown to be equal. Since then owing to

the confusion of the engraving [of the lines] he was not able to make the engravings of the hours in that [part] corresponding to the hemisphere above the earth, and consequently made them in the opposite one, for this reason he takes the degree of the sun according to diameter, and seeks how far this has been moved beneath the earth from the setting horizon, and declares that the movement of the sun above the earth from the rising horizon is equally great. This then is the reason of taking the degree of the sun according to diameter, whereby too the counting of the hours from the setting takes place under the hemisphere beneath the earth. In order that we may know exactly the amount of the portion of the hour when the degree of the sun according to diameter does not fall actually on the line of those marking the hours, but in an interval, we must mark with a dot the place where it fell, then having placed on the sign itself a reed moistened with ink, and kept it unmoved in the degree of the Arachne we have taken, and bringing it round together on the Arachne on each side as far as the lines marking the hours on each side, measure the whole line made by the ink on the tablet by a small cord or something of the kind, and then seek how great a part it is of this whole line up to the sign (i.e. the mark) on which fell the degree according to diameter of the observed (? degree). Thus the fraction of the hour, and its amount is declared.

6. *Another Method.*

We can discover the fraction of the hour in another and more technical way. It is necessary to watch how many parallels one of the indices ἐν τῷ μοιρογνώμονι of the Arachne passes through, wholly or in part, while the selected degree of the zodiac passes through the whole distance between the lines marking the hours on either side, on which it has fallen. Then again to look above how many parallels the same index passes through, or what part of one, while again the same degree passes through the sought-for part of the hour, as far as the sign between on which it has fallen, and so find out the relation (λόγον) of the part to the whole. Thus, for example, if the whole distance of the degree (μοιριαία διάστασις) passed through by the index is, let us assume, four parallels and a half, say that the part sought for is the third part of the hour. This it is possible to do on the instruments on which the outer edge of the tablets, or of the mater (δοχεῖον) itself that has been divided, has been divided into 360 sections of a degree by the index of the Arachne falling upon them. For having counted up how many degrees in the whole hour sought for the index passes through, and then again how many the same index passes through in the completed part of the same hour, what we seek, what amount it is of the whole, by the comparison of the measurements of these with one another, of those of the whole hour, and those of the part, we shall see how large a part of the whole is that which we are seeking.

7. *That four Central Points appear together, the Ascendant, the Culminating, and those Diametrically Opposed to them; and that it is possible in certain Instruments to take an Observation in any Tablet.*

From this placing we have at once the four central points, the Ascendant, the Culminating, and those Diametrically Opposed to them, I mean the setting one, and that beneath the earth to the culminating one. For the degree of the zodiac in which the sun is (here assumed to be the 20th of Aries) lying in the parallel in which it has been found (say in the 30th from the Rising), it is necessary to look which is the rising sign of the zodiac, and what numbered one of its degrees, or what part of one, touches the horizon, and say that this is the ascendant. Similarly it is necessary to see which sign of the zodiac is setting, and what degree thereof touches the setting horizon, that is to say the last parallel towards the west, and say that this is the setting central point. It is plain that the degree diametrically opposed to the rising one is the setting one. Further we must see what sign of the zodiac, and what numbered degree of it it is that touches the line corresponding to the meridian in the part of the tablet corresponding to the hemisphere above the earth, and say that these (keeping ταῦτα) are the culminating central point, and that diametrically opposed to it, in of course the culminating point, which in the remaining part of the tablet, which corresponds to the hemisphere beneath the earth, will fall on the line corresponding to the mid-heaven. And it is necessary to know this that in those (instruments) in which the outer edge of the tablets has been divided into the 360 degrees it makes |71 no difference to which tablet we adjust the rule and take the observation, the little gnomon itself falling upon them (sc. the degrees). The rest of the using we must do, as we have already said, adjusting the Arachne to the zone we are seeking.

8. *Concerning the skilful Observation of the Fixed Stars by Night.*

So much for observation by day. Now concerning that by night let us state that certain of the fixed and bright stars are engraved on the Arachne, in some cases seventeen, and in others more. Some of these of necessity in every night and every hour appear above the earth, as, let us say, *Lyra, Arcturus,* and others which you will find engraved on the Arachne. Beside each lies the special index of the star inscribed thereon. It is necessary then if we wish to take the hours in the night to observe one of the stars lying on the Arachne which appear above the earth. The method of observation will be as follows. We lift the instrument by the suspending ring, and place it above our eye, and turn the side of the instrument divided into the ninety degrees towards the star under observation, so that it lies as far as possible in the same plane surface as the star. Then, placing our eye beneath it, along the rule, we bring [the rule] gently round this way and that until the ray of the eye striking through the hole of the lower sight-vane to that of the upper, one beholds the star through both together. Here there is need of more exactness, lest having turned the eye aside we fail to note that we are looking at the star from outside, and not through the sight-vanes. It is therefore necessary to close one eye, and only observe with the other, lest any such error as we have mentioned should occur. Having then observed the star, we look at the degree on which the index of the rule fell, how far it is from the horizon, in the same way as was done in the case of the sun, and mark it. Then having sought the zone in which we were while observing, and the parallel therein in the same line with and of the same number as the degree under observation, we again mark it with ink. If then the star under observation happens to be in the quadrant of the meridian line, it is necessary to mark the parallel from the Rising, but if after the meridian from the Setting, like what happened in the case of the Sun. Then having adjusted the Arachne in the zone in which we were observing, we seek in it the star under observation, for instance let us say *Lyra* or *Spica Virginis,* or any other. This done we bring round the Arachne so that the index of the star may touch the parallel circle on which the star is being observed, and which also we marked. Then taking from the calendar the degree of the sun, in which it then happens to be, or by the method shortly to be described, we shall straightway find that it is in the semi-circle of the tablet on which the hours were engraved. Having marked it then with ink, and counted the hours from the Setting, and done all the other things as in the case of the sun, we shall find out the then completed hours of the night, or, if it so chance, the fraction [of an hour]. Similarly we shall see the four central points lying in their proper places.

9. *How we must know whether the Sun or each of the Fixed Stars when observed is before the Meridian Line, or on it, or after it:*

and how we can take the highest Altitude of each Degree of those in the Zodiac.

If then the star under observation, or the sun, is very far distant, either before or after the meridian line, it is easy to discern by perception which sections of the parallels we shall use, whether those before the meridian or those after it: for whether it has inclined far towards the rising or far towards the setting, in neither case does any mistake occur whether the star under observation or the sun is before or after the meridian. But if it is very near the mid-heaven, it will be uncertain whether it is before or after the meridian line. Which it is we distinguish in the following way: If it was the sun we were observing, we must seek what is its maximum altitude on the day on which we observed it. To know this we must take the sign of the zodiac and the degree in which the sun is on that day, as, let us say, the 20th of Aries, and having marked this degree on the Arachne with ink bring it round until it touches the meridian line, then seek on what numbered parallel it has fitted, and say that this is the maximum altitude of the sun from the earth being then in the 20th degree of Aries. This done, if the sun under observation is found on this degree, I mean of the maximum altitude, it is plain that it happens to be on mid-heaven itself, but if it has been observed to be of fewer degrees, it was before or after the meridian. To know this we shall wait a little, and observe again, and if we find the number of degrees increased, it is plain that it was before the meridian when first observed, if decreased, after it. And otherwise; having observed the sun we find it, let us say, elevated *seventy degrees above the horizon, and then are unable* to distinguish by perception whether it is before or after the meridian, we must, having waited again for a little, observe it once more, then if it has added, and become, say, seventy-one, it is manifest that the sun when previously observed was before the meridian, but if it has taken away, and become, let us say, sixty-nine, it is plain that when previously observed it was after the meridian. To know this we must count the parallels from the Setting, which first was observed lifted up from the horizon of the, as we have assumed, seventy degrees. We must then adjust the degree of the sign of the zodiac on the Arachne, in which the sun then happens to be, according to the parallel under observation, as in the present case the 70th from the Setting. If it is observed to be on the mid-heaven itself, it is plain that the degree of the sun will fall upon the line itself which corresponds to the meridian line, which also cuts the parallels; but if after the meridian, it will deviate from it in the direction of the Setting. So much in the case of the sun.

In the case of the stars we shall use the same methods seeking the maximum altitude of the star under observation in the zone in which it is being observed. This we shall know as follows. Having adjusted its index to the straight line corresponding to the mid-heaven, and looking on which parallel (it fits), if it fits on the line of the meridian line itself, we shall say that this is the maximum altitude in that zone, and then proceed as we directed in the case of the sun. Here too we should use the second method. We should again wait a little, and then observe, and so forth. For again if, having observed the star after a short interval, we find the number of degrees less, having fitted the index of the star under observation to the number of the parallel on which it was found to be in the first observation (counting the parallels, as I said, from the Setting) if we find the index of the star falling on the line itself corresponding to the meridian line, we say it was observed on the meridian line itself, but if it has deviated from it in the direction of the Setting, after the meridian line. From what has been said, it is plain how it is possible to obtain the maximum altitude in each zone of each degree of a sign of the zodiac. For, placing the Arachne on the tablet of the zone we are seeking, we must then bring round that degree of which we wish to obtain the altitude till it touches the meridian line, and straightway we shall find the inscription engraved of the altitude of the degrees. Thus then shall we be able to find the altitude of each degree. And this is to know the meridian line in each zone.

10. *How we can find out in how many Equal Hours each single Sign of the Zodiac is raised up, and in how many it sinks.*
We will add another use of the instrument, for we shall find out by its means in how many equal hours in each zone each of the signs of the zodiac is raised up from the rising horizon above the earth, and in how many it sinks down again. First it must be known that in the part of the instrument in which the tablets are introduced, and on which the Arachne is placed, which they were accustomed to call the mater of the tablets, there rises above it a certain rim divided, as I said before, into 360 degrees, which correspond to the segments of the equinoctial [circle], which they also call equal hours. When the whole instrument has been adjusted the rim, thus raised above it, becomes in a way continuous with the plane surface of the tablet lying upon it from outside, so that the whole plane surface becomes in a fashion one. For in complete instruments in which as a rule there is no mater, but each tablet is divided by itself on account of its size, and does not lie upon another, the rim aforesaid does not rise above it at all. But at each boundary of the tablet (? reading ἑκάστου, at the b. of each t.), that is to say on their circumference the aforesaid 360 equal hours are engraved, on which the index of the Arachne falls.

Example.
Let it be assumed that we are seeking in how many equal hours *Scorpio,* let us say, is raised up in the third zone. It is necessary then to place the Arachne on this zone, then to adjust the first degree of Scorpio to the first parallel from the Rising, and then to seek the index at the end of the Arachne ----it lies on the semi-circle outside the Arachne----on what degree of the said circle it has fitted, which [circle] we said

had been divided into 360 degrees, which are also called equal hours, and mark it. Then we must bring round the Arachne until the furthest degree of Scorpio, that is the 30th, has been brought up to and adjusted to the first parallel from the Rising. We must then seek again on which degree of the same circle the said index has fitted, and mark this too. We must next measure all the degrees from the original to the later mark, and as much of the 360 hours as we find the index has passed through in the whole raising up of Scorpio, say that in the same number of equal hours it is raised up. And so in each of the remaining cases. We shall then know in this way in how many equal hours each sign of the zodiac is raised up in each zone. In the same manner we shall find out in how many equal hours each sinks down, by similarly adjusting the first degree of the sign of the zodiac we are seeking to the furthest parallel towards the setting, and marking the degree on which the outer index of the Arachne fell. Then again bringing it round, and having placed its 30th degree to the same setting horizon, that is the furthest parallel, and having again looked where the index of the Arachne has fallen, and counted how many equal hours it has passed through in the whole descent of the sign of the zodiac, we shall say that in so many hours the sign of the zodiac descended below the earth.

11. *How we shall find out the Temporal Hour of each day and Night, and similarly of how many Equal Hours it consists.*

By the same method we can find out of how many equal hours each temporal day consists, and each hour. To know this we must again take the degree in which the sun is, and fit it to the first parallel from the Rising. Then mark the degree on which the index on the Arachne falls, and then bring round the Arachne until the degree of the sun is in the last parallel in the setting part, or, to say the same thing, until the sun passes out through the whole hemisphere above the earth. This done, we must again mark the degree, on which the index of the Arachne fell, and count the degrees beginning first from that which we marked up to the last, and say that the day in question is of so many equal hours. Then having divided these hours into 12, say of how many equal hours, or what part of one, each temporal hour consists. By the same method we shall find out the given temporal night and the length of its temporal hours, by placing the degree of the sun upon the setting horizon, that is on the furthest parallel, and marking on what degree of the outer edge of the instrument the index of the Arachne comes, then bringing round the Arachne until the degree of the sun having passed through the part of the tablet which corresponds to that beneath the earth touches the rising horizon, that is the furthest parallel towards the Rising. Having done this let us again look at the degree which the index of the Arachne touches. Then having counted them all from that marked in the first place we say that the temporal night in question is of so many equal hours, and having divided these into the 12 we find of how many equal hours the night hour consists. By this means then you have th_n dia&kristin [? the resolution] of the temporal into the equal hours.

12. *How we can find out by the Instrument the Epoch of the Sun. And herein again how we can get the Maximum Altitude of the Sun on each day.*

We can get the epoch of the sun without calculation by using the instrument in the following way. We must take the maximum altitude of the sun from the earth on that day. This we shall get by observing it at the meridian itself. It is plain that we must observe it very often until it no longer adds to its height, but, having been raised to its maximum altitude, begins again to grow less and come nearer to the earth. For it is clear that the point at which it began to diminish is its maximum altitude. Having got this we next look which quadrant the sun is passing through, whether it is that from the spring or that from the autumn equinox, or that from the summer or that from the winter tropics. (This is quite clear, for the hours, both the equal and the tropical, are known to all.) We shall then take this quadrant in the zodiac on the Arachne, and having placed the Arachne itself in the zone in which we are taking the observation, and fitted each degree of the quadrant, which the sun is then passing through to the mid-heaven, we shall seek which one of them is exalted so many parallels in the mid-heaven, and that which has become as great as the sun has been found to be exalted on that day, that we say is the epoch of the sun then. (ἐκείνην ἀποφαινόμεθα ἐπέχειν τότε τὸν ἥλιον). This happens if the sun is not near the tropics, but is far distant from them. If it is near we shall need another method of determination (διακρίσεως), which we will explain.

13. *Which Degrees of those in the Zodiac are under the same Parallels and are raised to the same Altitude, and how we can find in which Quadrant of the Zodiac the Sun itself is after the Tropical Signs.*

No degree of those in the same quadrant is raised to the same altitude as any other. In all the zodiac after the tropical signs you will find them raised to the same altitude only in pairs. These are those which are under the same parallel. Those under the same parallel are those distant by an equal interval from the tropical signs, separately from each----I mean from the summer one and the winter one, which are properly tropical. For from them the sun turns to the north and to the south: for from Capricorn it begins to be exalted towards the north as far as Cancer, and from there again begins to step back and go down to the south as far as Capricorn. Some people, however, say that the equinoctial signs are tropical, and so people in general say there are four tropics, owing to the changes of the hours. Properly then only the two signs are tropical----I mean the first degree of, say, Cancer, and the first degree of Capricorn. For now we must not be over-exact about these matters, which are not even co-ordinate with others. For no degree of the zodiac is raised to the same altitude as these. For there are boundaries of the obliquity of the zodiac. Consequently they are not even under the same parallel either with one another, or with any other degree of those in the zodiac. These distant by an equal separation from one of these on either side are under one and the same parallel, and consequently are raised to the same altitude above the earth. For example, the beginning of Leo and that of Gemini are distant by an equal separation on either side from that of Cancer, for there are thirty degrees on either side. These two degrees, then, the beginning of Leo and that of Gemini, are under the same parallel, and therefore are raised to the same maximum altitude above the earth. That what has been said may become clear, it is necessary to fit to the two boundaries of the hemisphere above the earth the two equinoctial signs of the zodiac, to the rising, let us say, the beginning of Libra, to the setting that of Aries. For these you will see having one and the same parallel, namely the first, by which the part of the tablet which corresponds to the hemisphere above the earth is marked off, and that beneath the earth. These then lying thus, you will see the first degree of Cancer, and the first of Capricorn, coinciding with the line of the meridian line, and in the direction of the equinoctial (ἐπὶ τὰ ἰσημερινὰ) is equally distant from the summer (τοῦ θερινοῦ)----I mean the first degree of Cancer [the first degree of Leo and the first of Gemini (this from the note)]. Consequently they are under the same parallel, as I said, and are raised to the same altitude above the earth. Then in the same way you will see again the remaining degrees, equidistant on either side from the first degree of Cancer, touching the same parallel, and raised to the same maximum distance. And these same (degrees) you will see equally distant from the two equinoctial signs. For things that are equidistant from one of the tropicals are also equidistant from both the equinoctials, either from other, but from one in the direction of what precedes (ἐπὶ τὰ ἡγούμενα), from the other in that of what follows (ἐπὶ τὰ ἑπόμενα). For example, as far as the beginning of Gemini is distant from that of Aries in the direction of what precedes, so far is the beginning of Leo from that of Libra in that of what follows. And again, as far as in the direction of what follows the beginning of Gemini is distant from that of Cancer, so far in the direction of what precedes is the beginning of Leo from that of Cancer. But it is not because they are equidistant from the equinoctials that they are under the same parallel, but because they are so from the tropicals. The beginning of Pisces and that of Taurus are equidistant from that of Aries, but neither are they under the same parallel, nor are they both raised to the same maximum altitude. For Pisces is more to the south, and Taurus more to the north. And further, as far as the beginning of Pisces is distant in the direction of what follows from that of Aries, so far is the beginning of Virgo distant in the direction of what precedes from that of Libra. But they are not under the same parallel, since Virgo is in the north, but Pisces is in the south. Since then those on either side equidistant from one of the tropicals are under the same parallel, and both the equinoctials are equidistant from the tropicals, and under the same parallel, therefore those on either side of the two equinoctials, equidistant either from other, the one in the direction of what precedes, the other in the direction of what follows, are in the same parallel. It will make no difference even if you place the beginning of Aries on the rising and that of Libra on the setting horizon, of course when that of Capricorn is culminating, for you will see the same things happening again. When only two of the signs are on either side of the tropicals being exalted to the same distance, if we seek the epoch of the sun when it is at the tropicals themselves, it will be hard to find out in which quadrant it happens to be, owing to the fact that the (beginnings) on either side equidistant from the tropicals have the same altitude. For example, the beginning of Cancer, say, being exalted Z degrees, and these on either side after ten, that is the tenth of Cancer and the twentieth of Gemini, as we have assumed, eighty-seven degrees [εἰ if], the sun being about the twentieth of Gemini, or the tenth of Cancer, we shall seek the epoch of the sun. Then having taken its maximum altitude being, as we assumed, of eighty-seven degrees, we seek in what degree of one of these quadrants in the Arachne it is exalted to this maximum, and we shall find that the tenth of Cancer and the twentieth of Gemini make the same the maximum altitude: and since we cannot by perception distinguish exactly whether before the summer tropics the sun is in the twentieth of Gemini, or after them in the tenth of Aries. (For the same thing happens in this case, which happened about the meridian itself, when we were observing the sun. This, to take a case, happens if, having stayed long in a desert, we are quite ignorant of the month, or among a people who either count the months in a different way from ours, or do not count them at all.) Again we wait for one or two days, and then make a similar observation, and if we find that the sun has added to the altitude, it is plain that it was previously before the summer tropics, but if it has taken from it, after them. Thus, then, if the sun be near the summer or the winter tropics. If, however, the sun is far distant from the summer or the winter tropics in this direction or in that, one doubtful point will remain, of which quadrant we must seek a degree elevated to as great a maximum as that of the sun we are observing. If the search be made before the summer tropics, from Aries to the thirtieth degree of Gemini, or, to say the same thing, the first of Capricorn, we must seek which degree of these is elevated to as great a maximum from the horizon as the sun is then observed to be elevated: but if after the summer tropics, from the beginning of Cancer to the thirtieth degree of Virgo, or, to say the same thing, the beginning of Libra. In the same way, if before the winter tropics from the beginning of Libra to the thirtieth of Sagittarius, or to say the same thing, the beginning of Capricorn; if after them, from the beginning of Capricorn to the thirtieth of Pisces, or, to say the same thing, the beginning of Aries.

14. *How we shall find the Epoch of each of the Planets.*

It is possible to get the epochs of the remaining planets by means of the instrument, with exactness when they are in the actual middle of the signs of the zodiac, and more roughly if they deviate to the one side or the other, in the following way. We must first observe one of the fixed stars inserted in the Arachne by the method already given. Then having learned how many parallels it has then been exalted from the rising or setting horizon we must place the Arachne in the zone in which we are making our observation in a manner corresponding to the position of the whole: this is to fit the index of the fixed star under observation to the parallel in which it has been observed to be. Then we must observe again the planet we are seeking, and mark how many parallels it has been raised from the setting or rising horizon, and seek the parallel of equal number in the zone in question, and the section thereof towards the west or the east in which the planet has been found to be. Then we must seek which degree of the zodiac this section of the parallel comes upon, and say that that is the epoch of the planetary star then under observation. It stands to reason that, as the sun always moves along (τὴν ? a path) through the middle (τῶν μέσων), it happens that we get its epoch by exact observation, since it is always borne in it (αὐτῇ) in which we determine the epochs of the times (χρόνων) But in the case of the rest, since they are not borne always in that, but often make their movement oblique in regard to it, so that they are sometimes more to the north of them, and sometimes more to the south, whenever we observe them deviating from this (ταύτης), if we lead out to the zodiac the straight (glance) borne to them from the eye, it is necessary that it should fall not on that through the middle, but outside it, either more to the north, or more to the south, of it. Consequently we cannot get their epoch exactly, since, as I said, we can only determine the epochs by that through the middle.

15. *How we can discover how far each Degree of the Zodiac deviates from the Equinoctial to the north or the south. And similarly in the case of the Sun and the Moon, or each of the Planets.*

We shall find by using the instrument how far in width each degree of the zodiac is distant from the equinoctial in a northerly or southerly direction in this way. We have said in what preceded that the distance between the circle of Capricorn and that of Cancer has embraced the whole obliquity of the zodiac, amounting to forty-eight degrees, for from the latter to the equinoctial are twenty-four degrees, and from the equinoctial to the former the remaining twenty-four. It is plain that from the winter to the summer tropics the sun passing through the whole semi-circle is exalted towards the north, and reversely from the summer to the winter ones is lowered towards the south, and it is obvious that between the circle of Cancer and that of Capricorn is the equinoctial circle. Hence it comes about that twice in the year the sun is in this, from the summer tropics to the winter coming by Libra, and from the winter to the summer by Aries, with the result that in each semi-circle the sun is at one time to the north, and at another to the south, of the equinoctial. If then we wish to find how far each degree of the zodiac in either semi-circle is distant to the north or to the south from the equinoctial circle, we shall find it in this way. Of the equinoctial signs----I mean the beginning of Aries or Libra---- we must fit to the mid-heaven above the earth, and mark the parallel on which it falls: then again fit the degree of the zodiac we are seeking to the same meridian line, and mark the parallel on which it has fallen. This done, as many parallel circles as we find from the equinoctial to that degree, so many degrees we shall say it is distant from the equinoctial, and whether to the north or south we have at once by inspection what is being sought. For if the degree we are seeking falls outside the equinoctial, as in the direction of the winter tropical, as has been engraved on the tablet, it is plain that it is distant from the equinoctial in a southerly direction: but if within the equinoctial, as in the direction of the summer tropical, which we say is contained by the equinoctial, it is plain again that the degree of the zodiac we are seeking deviates in a northerly direction. And this is plain from the mere placing of the signs of the zodiac. For if we seek the (τὰ) after the beginning of Aries up to the twenty-ninth of Capricorn, it is plain that they deviate to the north of the equinoctial, and if those after the beginning of Libra up to the twenty-ninth of Pisces they will have a deviation to the south of the same mid-heaven. So it is plain to everyone, that for the sun, the moon, and each of the planetary stars, taking them in each occurring degree of the zodiac, we thus arrive at how much they deviate to the south or to the north of the equinoctial. For having taken the degree, in which the sun, or the moon, or each of the planetary stars, has its epoch, and done all the things previously mentioned, we shall find what we were seeking. For by as much as the degree of the zodiac deviates from the equinoctial to the north or to the south, so great a deviation does the star upon it make. Using the same method we shall know whether each of the fixed stars in the Arachne, is more to the south or more to the north of the equinoctial, and by how many degrees it is distant from it in either direction.

Made in the USA
Monee, IL
13 September 2026

62976954R00017